T0003961

Time

10 Things You Should Know

A brief journey through the greatest mystery of our universe

COLIN STUART

SEVEN DIALS

For Arthur, how lovely to have finally reached the region of spacetime with you in it

and for my great-great-grandchildren: come and visit, I'll put the kettle on . . .

First published in Great Britain in 2021 by Seven Dials
an imprint of The Orion Publishing Group Ltd
Carmelite House, 50 Victoria Embankment
London EC4Y 0DZ

An Hachette UK Company

3 5 7 9 10 8 6 4

A CIP catalogue record for this book is available from the British Library.

ISBN (Hardback) 978 1 8418 8492 9
ISBN (eBook) 978 1 8418 8493 6

Printed in Great Britain by Clays Ltd, Elcograf S.p.A.

www.orionbooks.co.uk

Contents

Preface

We tie it to our wrists and hang it on our walls. We mark its passage with candles and fireworks as it in turn marks our faces with lines and wrinkles. It can be wasted and killed. Spent and saved. Kept and lost.

We're so obsessed with time that it is the most used noun in the English language. Time will tell, waits for no one, will mend a broken heart and flies when we're having fun. More often than not we wish we had more of it. How often, though, do you stop and think about time itself? Could you explain it to someone else? The moment you do, it feels like trying to hold a puddle of water in your hands. It seems to wriggle through the gaps.

The Roman philosopher Augustine of Hippo, known more popularly as St Augustine, perhaps put it best when he said,

'What is time? If no one asks me, I know; if I wish to explain [it], I know not.'

The book you're holding is my attempt to explain time from the point of view of a physicist. As you'll see, time is not just one of the most enduring mysteries in science, but it is one of the most enduring mysteries in *all* human experience. We'll embark on a journey from when we first started keeping track of time thousands of years ago to the forefront of modern physics research. We'll see how time can be slowed down, sped up and maybe even stopped. You'll discover how to travel back in time to meet your past self and perhaps even discover that time may not exist at all.

The celebrated physicist Richard Feynman once compared trying to decipher the laws of physics to playing a game of chess; except that you can't see all the pieces at once and no one has told you the rules. You have to figure them out for yourself through a combination of experience, experiment and occasional glimpses at the board. Up until the last few centuries, we humans spent most of our multi-millennia existence living in a tiny corner of the board, seeing only a very limited number of moves play out. This has led us to think about time in a very particular way and those ideas have become firmly entrenched. They're wrong. Physicists have shown beyond a shadow of a doubt that time does not work in the way you think it does.

Think back to your distant ancestors. It would have been very rare for them to leave their home town. Rarer still to journey abroad. When you can travel outside the confines of your immediate existence you realise there's more to the world than meets the eye. Cultures, languages and ideas that you'd never conceived of before. Wonders untold. However, when it comes to time, few of us have ever left our little mental hamlet. The incredible developments in physics over the last century or so – since Albert Einstein gifted the world with his theories of relativity – provide a more complete picture of time.

Reading about these ideas as a teenager changed my life. After discovering them I was never going to study anything other than physics. I simply had to know more. At a stroke I realised the subject was far richer than the boring circuits and levers taught in the classroom. I've spent the last twelve years writing and speaking about time and space because I want to give you that same revelatory moment. I've seen jaws drop, but I've also seen how reluctant people are to give up their cherished notions of time.

Encountering these ideas for the first time is a bit like a deep-tissue massage. It'll make you uncomfortable at first, but in the end it's so worth it. Cast aside what you think you know about time and you may just fall in love with physics as much as I have.

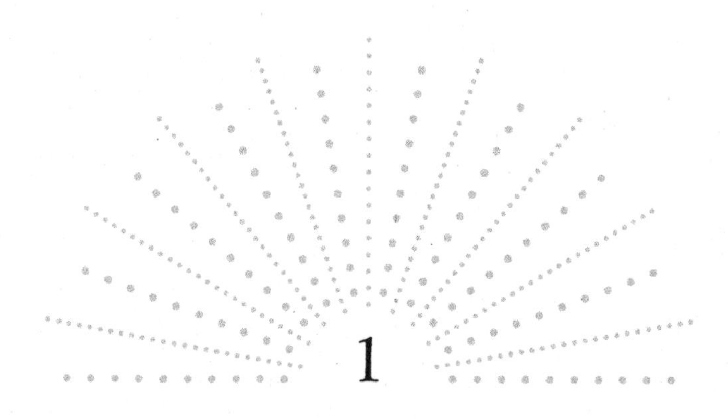

1

The Earth Is a Terrible Timekeeper

American swimmer, Nathan Adrian can hear the muffled cheering of the crowd through the water. He's giving it everything to reach the other end of the pool first. In the next lane is the race favourite, Australian, James 'The Missile' Magnussen. The two men are neck and neck and appear to touch the wall together, before the result confirms that Adrian is the victor by a margin of just 0.01 seconds.

This dramatic end to the 2012 Olympic 100 metre (328 feet) men's freestyle final underscores the extent to which we increasingly carve up time into miniscule amounts. The tiniest fraction of a second can be the difference between gold and silver. It can also make you a fortune. In 2009, construction workers installed a 1,300-kilometre (807 mile) -long underground cable between the stock exchanges in Chicago and New York at a cost of $180 million. All to shave 0.000004 seconds off the time it takes to send trading information between the two hubs. Even that small difference increased profits by £12 billion a year in an industry where time is quite literally money.

When our ancestors first decided to break up the

unceasing passage of time into manageable chunks they had no need for such tiny divisions. In fact, the first minute hand didn't appear on a clock until 1680, with the second hand following a decade later. Their system of time was based on the sky. The gap between sunrises became a day, which was divided into hours, minutes and seconds (the last two were originally called 'first minutes' and 'second minutes', which is why we now call them seconds). They collected days together into blocks of seven, called weeks – one day for each of the seven celestial objects they could see moving in the sky (SATURN-day, SUN-day, MOON-day, etc.).[1] A month – or MOONth – is how long it takes for the Moon to wax and wane through its cycle of phases.

---

1 The other planets aren't apparent in English due to the influence of Norse mythology (Thursday being Thor's Day, for example).

To see the planets, switch to a Latin language instead – like French.

Tuesday = Mardi (Mars)
Wednesday = Mercredi (Mercury)
Thursday = Jeudi (Jupiter)
Friday = Vendredi (Venus)

The pattern of the seasons repeats after 365 days – a year – which today we know is because that's how long it takes for the Earth to orbit the Sun. So each birthday is really a celebration of completing yet another billion-kilometre (621-million-mile) -long lap of our nearest star.

This ancient system creaks under the weight of our modern digital age, particularly as our planet is a terrible timekeeper. At the heart of the issue is the variability of Earth's spin. The day lasts twenty-four hours because that's how long it takes for the Earth to rotate once on its axis so that the Sun returns to the same spot in the sky. Yet, far from being fixed, Earth's rotation speed can change.

Back in 2011, a magnitude 9.1 earthquake struck in the Pacific. It created a colossal tidal wave that swelled to a height of 40 metres (131 feet) and crashed over a huge area of Japan, killing more than ten thousand people. Over a hundred thousand buildings collapsed. At the Fukushima power plant three reactors went into meltdown in the worst nuclear accident since Chernobyl. The quake was mighty enough to move the entire island of Honshu – Japan's largest – by over 2 metres (7 feet). It also made the day shorter. The force of the earthquake sped up

Earth's rotation, slicing 1.8 millionths of a second off the length of the day. Earthquakes in Chile in 2010 and Sumatra in 2004 reduced the day by roughly similar amounts.

The Moon's gravity also wreaks havoc with timekeeping. A vast mound of water rises on the side of our planet nearest the Moon as our neighbour tugs on us. Anyone on the coast in this location experiences a high tide. The Moon does its level best to keep this tidal bulge in place, but the Earth is turning underneath it. If you're standing at the beach and see the tide go out, you'd swear it was the sea running away from the sand. In reality, it is the Earth's rotation carrying you and the shoreline away from the water. The Earth loses some of its rotation speed as it struggles to spin underneath the tidal bulge, making every day a little longer in the process. This gradual but relentless effect is currently extending the day by 0.0017 seconds every century. Although that doesn't sound like much, it all adds up. Incredibly, 430 million years ago, the day was just under 21 hours long and there were 420 sunrises a year instead of 365. This isn't simply a guess: fossilised coral seals the deal. Coral is made of calcium carbonate, which is laid down in lines each day as the organism grows. Corals grow more in the dry season than the wet season, meaning the lines are arranged into a pattern that repeats every year. Each block contains 420 lines, corresponding to a year lasting 420 days.

Earth's lousiness as a clock has profound consequences for how we define the second. There are 86,400 seconds in 24 hours and so the length of a second used to be defined simply as 1/86400th of a day. Yet if the length of the day is variable, so is the duration of a second. How can a second today be shorter or longer than a second yesterday? As computers became more prevalent, it was essential that they all operated with a consistent system of time. We needed to standardise the second. So, in 1956, scientists redefined the second as a fraction of the year rather than the day because the Earth's rotation around the Sun is more reliable than its spin on its axis. That made the second equal to 1⁄31,556,925.9747th of a year. Not just any year, though. Very specifically the duration of the year 1900. Even the length of a year varies slightly, meaning they had to pick one. This change turned out to be temporary. Eleven years later, in 1967, we moved away from using the unreliable Earth as a timepiece entirely. The second is now defined using the building blocks of the universe: atoms.

Atoms are tiny. There are more atoms in a teaspoon of water than there are teaspoons of water in the Atlantic Ocean. Perhaps the simplest way to picture an atom is as a mini version of the solar system. There is a central nucleus, which is a bit like the Sun. Then there are particles called electrons whizzing around it, akin to planets. These electrons are restricted to only occupying certain orbits, which physicists call energy levels. The lowest of these energy levels – the tightest orbit around the nucleus – is called the ground state. Energy levels are often spilt into several distinct layers, which physicists refer to as fine structure and hyperfine structure. When an electron drops to a lower energy level it gives out the energy it loses in the form of radiation.

Putting all that together, we arrive at the modern definition of the second as adopted by the International Committee of Weights and Measures at their thirteenth official meeting in 1967.

Brace yourself, it's one hell of a mouthful. 'The second is the duration of 9,192,631,770 periods of the radiation corresponding to the transition between the two hyperfine levels of the ground state of the caesium-133 atom.'[2] The wording of the definition was tweaked again in 2019, although the details remain the same.

Thankfully this all went on behind the scenes, with the general public largely unaware of the change. However, it does occasionally rear its head in a way you may have noticed: leap seconds. They stem from the fact that we essentially now have two systems of time – atomic time and solar time. Atomic clocks based on caesium can tick for a few hundred million years before they lose a second of time, so the two systems gradually drift apart as the Earth's inability to keep good time becomes apparent. This is why every now and then we have to make a calendar year one second longer to bring the two systems back into agreement. The final minutes of 2005, 2008 and 2016 all lasted 61 seconds. At the time of writing, 27 leap seconds have been added since the first in 1972. That year ended up being 2 seconds longer, with a leap second added in both June and December. If we let the

---

2 https://www.bipm.org/metrology/time-frequency/units.html (accessed 02/10/2020)

two systems drift unchecked, we would eventually arrive at the ludicrous situation where our clocks say noon even though it's the middle of the night.

It's a system we're already used to with leap years. The modern calendar year is 365 days long, but it takes an average of 365.2425 days for the Earth to journey around the Sun. Without adding an extra day to February every four years to account for the missed quarter days, the seasons would drift and we'd eventually end up with winter in June and summer in December in the Northern Hemisphere. This system is called the Gregorian calendar after Pope Gregory XIII, who introduced it in 1582. Before that, many European countries relied on Julius Caesar's Julian calendar, which didn't account for the additional quarters. Britain eventually switched over to the new calendar in 1751 to stop the seasons drifting any further, which brought an abrupt end to that year after just 282 days. Then, to sync up time with the rest of Europe and undo the existing drift, eleven days were sliced from 1752. Wednesday, 2 September was followed by Thursday, 14 September. Heaven forbid if you had a birthday in between.

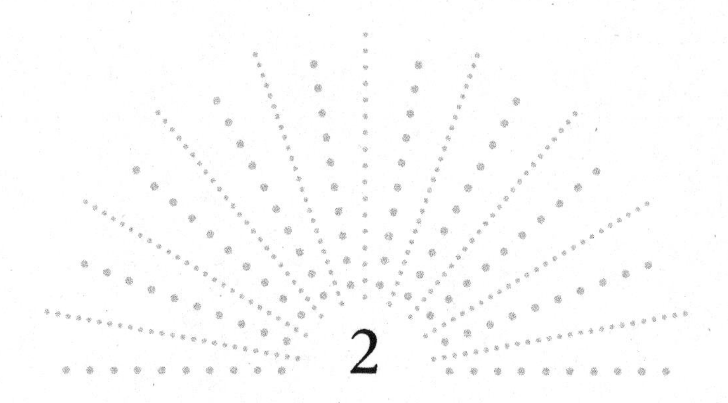

2

Rocks Are Clocks

Archaeologists are hard at work in the dark, dank caves of Bulgaria's Dryanovo river, scouring every centimetre for clues of the humans that were here before them. They hit the jackpot when they find a tooth. Looking even closer, they find six human bone fragments, pendants made from bear teeth and animal bones showing clear signs of butchery.

The big question is: how long have they been there? Our relationship with the past is a tentative one. The longer we live, the sketchier our memories become. It's why we've invented better ways of preserving history. The first film was recorded in 1888, a 2.1-second-long snippet of Roundhay Park in Leeds, England. A year later, Benjamin Harrison became the first US president to have his voice recorded in a thirty-six second clip. The first photograph takes us back to 1826, when Joseph Nicéphore Niépce snapped the view out of an upstairs window at his home in Burgundy, France.

Before these revolutionary inventions, we relied on the written word to keep track of the passing years and the important events within them. Even that only takes us back so far. Cuneiform script – widely recognised as the world's first writing system – was devised in Mesopotamia (modern-day Iraq) a little over five thousand years ago. What of the world before we could write? When did humans first appear? How long had the Earth been around before we popped up? It's exactly these enduring and engaging questions that inspired archaeologists to go trawling for evidence in the Bulgarian caves.

Back in the lab, after careful analysis, they estimated the tooth to be between 44,000 and 46,000 years old. Their results, published in May 2020, represent the oldest human remains ever found in Europe. They help paint a picture of when our early ancestors first migrated out of Africa and began to spread across the world's sprawling continents. But how did they know how long the tooth had been sitting there? Fortunately nature has its own clock – one that has been steadily ticking ever since that tooth first left the gums of its owner. It's called radioactivity.

To understand radioactivity, we need to dive deeper inside the atom than we did in the last chapter. A quick reminder: an atom contains electrons orbiting a central nucleus. That nucleus is made of particles called protons and neutrons. All atoms crave stability, but some atoms are heavy and unstable because they have too many neutrons. Let's take carbon as an example. Most of the carbon in the world is called carbon-12. The number here tells you the total number of particles in its nucleus (6 protons plus 6 neutrons). Carbon-12 is stable, but there's a far rarer form of the element called carbon-14 with 2 additional neutrons. That makes it unstable. It can stabilise itself by turning one of its neutrons into a proton, metamorphosing – or decaying – into nitrogen-14. Scientists refer to the original nucleus as the parent nucleus and the new one as the daughter nucleus.

This radioactive process is entirely random. It's impossible to predict when any individual atom of carbon-14 will shape-shift in this way. However, let's say you have 1 gram of carbon-14. It contains a staggering 43,000 million million million carbon-14 atoms. From what we know we are able to say how long it will take for half of these atoms to turn into nitrogen-14; we just don't know which half. The answer is 5,730 years and this number is called the half-life. Every 5,730 years, the number of carbon-14 atoms in a sample will drop by half. After 11,460 years – two half-lives – there will be a quarter of the original amount remaining. Another half-life later there will be just an eighth left. By carefully measuring the abundance of carbon-14 in the tooth from the Bulgarian cave, the archaeologists were able to peg it at between 44,000 and 46,000 years old. This immensely powerful technique is called radiocarbon dating.

Just like photographs, video recordings and writing, radiocarbon dating can only take us back in time so far. Venture too far into the past and there will be so little carbon-14 left that it renders the technique useless. It is also only good for living things, which had a high level of carbon to start with. If we want to work out the age of older objects – even the Earth itself – then we must turn to atoms with clocks that tick much more slowly. Those with much longer half-lives. For that we need the element uranium.

The nucleus of a uranium-238 atom is unstable because it contains 92 protons and 146 neutrons. By shedding 2 protons and 2 neutrons it decays into thorium-234. Thorium-234 is also unstable and so it too decays. A cascade of radioactive decay continues until a stable daughter nucleus is reached, which in this case is lead-208. The half-life of this radioactive chain is around 4.5 billion years. We can use this to estimate the ages of rocks on Earth in just the same way as carbon-14 is used to date biological samples. This time it is called radiometric dating. The amount of lead-208 in the oldest rocks on Earth suggests that one half-life has passed since they formed. Earth is a whopping 4.54 billion years old. This is backed up by the analysis of meteorites – space rocks that have fallen to Earth. Often they're chipped off pieces from asteroids, lumps of rock and metal that roam the solar system. Asteroids are building blocks left over from the formation of the planets. Radiometric dating of meteorite material confirms they are of a similar age to the Earth.

It's easy to throw around big numbers like 4.54 billion years without actually stopping to think about just what a huge stretch of time that is. If you reduce each year of Earth's existence to a second, you'd still have to wait 144 years – roughly two lifetimes – to get through them all. Collect a penny for every year that has elapsed and you'd not only be a multimillionaire, your pile of metal would weigh more than the Eiffel Tower. Humans have been around for so little of the Earth's history, roughly the last 300,000 years. Condense the Earth's

4.54-billion-year lifespan into one twenty-four-hour period and the first modern humans appear at about six seconds to midnight. If the same stretch of time is represented by a 1-metre (3 feet) -long stick, all of human history would be crammed into the last 0.07 millimetres (0.003 inches). That's less than the width of a human hair and a distance too small for the human eye to discern. We're mere flotsam and jetsam on the wide, open ocean of history.

Yet even the Earth is a youngster in comparison to the universe as a whole. Astronomers can't go digging for clues like archaeologists, but that hasn't stopped them from finding their own celestial fossils: globular clusters. You can see them for yourself if you run a pair of binoculars around the night sky. They look like a swarm of fireflies, a hive of stars all buzzing through space together. Astronomers can tell how old a star is by what it is made of. When the first stars formed, they only contained hydrogen and helium. Yet stars are the ultimate element factories, churning hydrogen and helium into the smorgasbord of other elements you'll find in the Periodic Table. When stars die, they release these heavier elements into the universe and they eventually end up inside new stars. This means that stars formed relatively recently contain lots of different elements, whereas the oldest are more pristine. In July 2020, astronomers used globular clusters to estimate that the universe is more than 13 billion years old – almost three times the age of the Earth.

This ties up nicely with another, independent way of measuring the universe's age. Globular clusters are normally found orbiting galaxies – huge stellar cities containing several hundred billion stars. Our own galaxy is called the Milky Way and it's just one of an estimated two trillion galaxies in the universe. We've known for nearly a century that almost every galaxy appears to be moving away from the Milky Way. The universe is getting bigger day by day. That means it must have been smaller yesterday and smaller still back when those early Europeans were making jewellery out of bear teeth in the Bulgarian caves. Astronomers can keep winding the clock back to see when the expansion started and everything in the modern universe was concentrated into the same spot before flying outwards. The fabled Big Bang – the event that created all time and space. According to the latest astronomical measurements the Big Bang happened 13.8 billion years ago.

We've arrived at time zero, the earliest point in

history. Before these discoveries, many astronomers thought that the universe had been around forever, pretty much in its current form. Now we know that time had a beginning and that the universe has existed without the Earth or humans for most of its life.

3

Telescopes Are Time Machines

Far from the bright lights of a city, the stars shine down with fierce intensity. Planets blaze enough to cast shadows and the dusty Milky Way arches overhead like a beautiful cosmic rainbow. Although it may not seem like it, this stunning skyscape is a type of time machine. One that allows you a glimpse of the distant past, if not the chance to visit in person.

Nothing in the universe can travel as fast as light. It zips across space at the remarkable pace of almost 300,000 kilometres (186,411 miles) per second. That's swift enough to whip around the Earth 7.5 times in the time it takes for your heart to beat once. This dizzying velocity means that as we go about our daily lives, light might as well travel instantaneously. Flick a switch and there's no hanging around waiting for the light to reach your eyes. The actual delay – a billionth of a second for every 0.3 metres (1 foot) you are from the bulb – goes unnoticed.

This changes significantly as soon as you start venturing into space, and the distances light has to travel literally become astronomical. The Moon – our nearest celestial neighbour – sits 384,400 kilometres (238,855 miles) away. Light takes 1.3 seconds to cover that distance, so we say that the Moon is 1.3 light seconds away. Think of light as being like a postcard sent across space. A postcard never tells you what the writer is up to when it drops on your doormat. Instead it tells you what they *were* doing when they wrote the card several days earlier – it takes time for the message to travel to you. Postcards and light always show you past events. It means that you never see the Moon as it is now, but as it was 1.3 seconds ago when the light that's entering your eyes left the Moon. This is true of your daily interactions with people too. Stand a metre (3 feet) away from someone and you're not seeing them now, but as they were 3 billionths of a second ago when the light left their face. The fact that it takes time for light to travel means there isn't really a consistent 'now' at all. We are always seeing the past, never the present.

By the time you reach the familiar stars of the night sky, the distances have grown from light seconds to light *years*. The nearest star to us after the Sun – Proxima Centauri – is 4.2 light years away. It takes 4.2 years for light to cover the 40-trillion-kilometre (25 trillion mile) distance between us. People often get confused with light years, naturally assuming from the name that they must be a measure of time. In fact, light years measure distance. One light year is the distance light travels in one year, so 9.46 trillion kilometres (5.88 trillion miles). (Using them saves on a lot of zeros.)

One of the brightest stars in the night sky is Betelgeuse, a red jewel some 700 light years away in the constellation, Orion. Betelgeuse is reaching the end of its life and will eventually detonate in a cataclysmic explosion called a supernova. For a time it will outshine the full moon and even be visible during the day. The thing is, Betelgeuse could have already exploded. If it blew up 699 years ago then the light from the supernova wouldn't reach us until next year. We're seeing a past version of Betelgeuse in the night sky, not a current one. It may even already be dead.

A fun thing to try is to look up a star online that's the same number of light years away as your age[3]. So if you're thirty-seven, you're looking for a

---

3 If you're forty-nine years or under, there's a handy list here: http://en.wikipedia.org/wiki/List_of_nearest_bright_stars (accessed 02/10/20). But you can also find other resources online.

star that's thirty-seven light years away. Arcturus (36.7 light years away) is a pretty good fit here. It is a bright red-orange star not far from the group of stars known as the Plough or Big Dipper. Find it in the night sky and the light entering your eyes now has been travelling across space ever since your eyes first opened thirty-seven years ago. After each birthday you'll have to change to a star one more light year away in order to keep up with the fact that you're another year older. In the gap between birthday cakes, starlight has been busy travelling another 9.46 trillion kilometres (5.88 trillion miles).

The furthest object you can easily see in the night sky is Andromeda, the nearest major galaxy to our Milky Way. To find it, first identify the prominent W-shaped constellation of Cassiopeia. If you picture each half of the W as a downward-pointing arrow, then the right-hand arrow points towards Andromeda. If you're somewhere with a dark sky that's free of light pollution, then you should be able to see it with your own eyes. Otherwise a pair of binoculars should do the trick. You're looking for a fuzzy, cotton-wool-like smudge.

It may not look that impressive but Andromeda is a mega stellar metropolis, home to a trillion stars.

You only see it as a faint smear of light because Andromeda is a whopping 2.5 *million* light years away. The light you're seeing first set off 2.5 million years ago, at a time when one of our ancestors – *Australopithecus* – was just starting to fashion the first tools out of stone. The first human species – *Homo habilis* – had appeared just 300,000 years earlier. Almost all of human history has played out in the time it's taken for the light from Andromeda to reach us. Gazing out into space is a way to look a long way back in time.

Let's flip things round and imagine a planet in Andromeda that's home to an advanced civilisation of aliens who are also capable of building telescopes. If they were able to zoom in and see the Earth, they wouldn't see a world of smartphones, social media and selfies. That light has only just left us and has a long way to go to get there. Instead they'd be receiving the light that set off from Earth 2.5 million years ago. They'd see a planet inhabited by *Australopithecus*, sabre-toothed tigers and giant mammoths. A civilisation in a galaxy 66 million light years away could drop in on the dinosaurs just before they went extinct. Our history is surging out across the universe at breakneck speed. Images of Hitler, Marie

Curie, Genghis Khan, your great-great-grandparents and you as a child are flooding through space, bringing our backstory to anyone with the tools to tune in.

However, as with all these things, there is a catch: you'd need one hell of a telescope to see any of it. Think of a telescope as a bucket and imagine that stars are firing out balls in all directions in the form of starlight. If you're close to the star, then you don't need a very big bucket to have a fair chance of capturing a good number of balls. The further away the star is, the bigger your bucket needs to be to see and photograph an object. Digital photographs are made from a patchwork of tiny squares called pixels. If an alien civilisation wants to capture the Earth as a single pixel in a photograph, then the mirror in their telescope needs to be 1 kilometre (0.62 miles) across for every light year they are away from us.

So, let's say that you're an alien on a planet a mere 150 light years away – a typical distance for a star in our night sky. The light currently reaching you from Earth would have departed back in the nineteenth century, during the reign of Queen Victoria. You'd need a telescope with a mirror 150 kilometres (93.2 miles) wide to see the Earth as a single pixel. One end would be in north London and the other in south Birmingham. For comparison, the biggest telescope mirror in the world right now is 10 *metres* (33 feet) across. Still, it's not impossible that an advanced alien race could build a telescope that big, either on the ground or in space.

If you wanted to see Queen Victoria herself, though, it's a different story. She stood 1.52 metres (5 feet) tall, meaning you'd need a telescope as wide as the distance between Earth and Saturn – 1.26 billion kilometres (783 million miles) – to capture her as a single pixel. For a more detailed photograph it would have to be bigger. Travel further back in time and the mirror grows larger still. A telescope capable of seeing a *T. rex* from 66 million light years away would span over six light years. That's more than the distance between the Sun to the next nearest star. A mirror that big – if it were made of glass – would collapse in on itself and form an object called a black hole (plenty more on those later in the book). Even if there was a civilisation living around our closest neighbour, Proxima Centauri, they would need a telescope nearly 3,000 times wider than the Earth to see you 4.2 years ago.

So our history *is* out there among the stars, but it's unlikely anyone will see it in any detail. We can continue to gaze up at the night sky – and back in time – without fear of an alien peeping Tom watching our every move from afar.

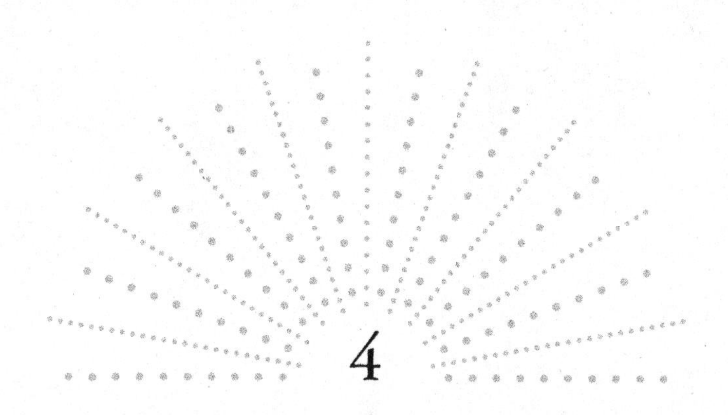

4

Time Flies Like an Arrow

Frenchman Sadi Carnot died in a Parisian asylum in 1832 at the age of just thirty-six. A victim of a cholera outbreak, the disease was so contagious that most of his writings shared his grave, for fear that they were also contaminated. The single book that did survive him would eventually help to explain one of the most commonly asked questions about time: why does it only run forward?

You don't need me to tell you that the future relentlessly becomes the past and never the other way round. We live from one moment to the next, not one moment to the last. We remember yesterday, but not next week. We're born before we die. British astronomer Arthur Eddington referred to this as time's arrow, one that points us ever towards tomorrow. Some punning wit later came up with: 'Time flies like an arrow; fruit flies like a banana.'

Curiously, despite being such an obvious feature of our everyday lives, the majority of the laws of physics don't care about a direction of time. Imagine watching a film of a ball flying in a straight line through the air. You'd be hard pushed to tell whether you were being shown the video in the order it was filmed or whether someone was playing it to you backwards. The equations describing its motion work equally well either way. Yet watch a video of a mug smashing on the floor and you'd know immediately if you were being shown it in reverse. Broken mugs don't spontaneously repair themselves. Cracked eggs don't re-form. People don't get younger.

It was the work of Sadi Carnot that set us on a path to understanding why time flies like an arrow in this way. The son of Napoleon Bonaparte's minister of war, Carnot was an engineer in the French army and dedicated much of his short and tragic life to understanding a newfangled invention that would soon revolutionise the world: the steam engine. His book, published while he was still in his twenties, was called *Reflections on the Motive Power of Fire.* It showed how to make steam engines far more efficient and, after his death, it inspired other physicists to found an entirely new discipline called thermodynamics. There are now four 'laws' of thermodynamics – a quartet of seemingly unbreakable rules we've discovered regarding the way heat and energy operate.

Today these laws have reached a sort of cult status,

one that sees them crop up in popular culture from time to time. In an episode of *The Simpsons*, Homer barks at Lisa, 'In this house we *obey* the laws of thermodynamics.' Netflix even released a rom-com in 2018 called *The Laws of Thermodynamics* in which a physicist attempts to use the laws to describe romance and relationships. The most important law for us here – and probably the most famous – is the second. In short, it says that disorder always increases. Anyone who's ever tried to keep a house, garden or desk tidy can vouch for this. Physicists have a special word for disorder. They call it entropy. According to the second law, entropy is always increasing and the universe gets messier and messier as a result. The arrow of time consistently points from a region of low entropy (the past) to one of high entropy (the future).

The reason that the universe gets more disordered over time is simply that it's by far the most likely outcome. To see why, let's take a set of six dice. Initially you set them up so that the top sides each show the same number. This is an incredibly ordered, low-entropy system. You then roll the dice. The likelihood of them still all showing the same number is just 0.013 per cent. You could roll the dice a hundred thousand times and it would only happen thirteen times. The other 99,987 times, the system would become more disordered and the entropy will have increased. That's just for a group of six objects each with six different possible configurations.

Imagine opening a brand-new pack of playing cards instead, with the deck neatly arranged in numerical and suit order. This is another highly ordered, low-entropy state. The chance of randomly shuffling the deck and returning it to this ordered state is 1 in 10,000 quadrillion quadrillion quadrillion quadrillion (1 followed by 64 zeros). To put that number into context, if I hid a solitary atom somewhere in the Milky Way – our home galaxy containing hundreds of billions of stars – you'd have the same odds of finding it by picking an atom at random. You're more likely to win the jackpot on the lottery eight times in a row.

The real world is far more complicated than a set of six dice or deck of fifty-two cards. Place an otherwise empty box measuring just a metre (3 feet) on each side in the middle of a room and it will contain 10 trillion trillion air molecules. As these molecules jive and jostle about they could become more ordered, perhaps by all gathering in one half of the box. Such a move from disorder to order – though not impossible – has odds so vanishingly small that we'd have to wait many, many times longer than the age of the universe to see it happen. So we never do. Order always deteriorates and it's this that gives us an arrow of time pointing from the past to the future.

This argument, based on probabilities, was first made in the 1870s by Austrian physicist and thermodynamics heavyweight Ludwig Boltzmann. A man whose story, like Carnot's, is tragic. After years battling both the scientific establishment and his own mental health, Boltzmann hanged himself while on holiday with his family in the summer of 1906. He left no note. Boltzmann is buried in the same cemetery in Vienna as Beethoven and Brahms, his tomb engraved with the equation that's still used to this day to calculate entropy.

Now you may be thinking that it is possible to make a system more orderly. You may clean your bedroom or organise your paperwork. Rather than relying on chance, you could spend time putting the dice or the cards back the way they were. Wouldn't you then have increased entropy? You would, but only locally. Tidying requires you to expend energy – you get hot and bothered clearing up a messy bedroom and probably make a fair amount of noise to boot. This energy escapes into the room and increases the entropy of the air around you, far outweighing the small local entropy decrease you've managed to achieve by finally being able to reach the bed without stumbling over your books. You've also made the world more disordered by breaking down food to give you the energy to tidy up in the first place. The overall entropy of the universe has still increased and there's no violation of the sacred second law of thermodynamics.

So what will happen to the universe if disorder keeps increasing? Eventually the cosmos will reach a state of maximum disorder. There will no longer be energy available to create brief pockets of order. No stars, no planets, no life. The good news is that the number of years it will take to arrive at this point is at least one followed by a hundred zeros. Astronomers call this the heat death of the universe (or sometimes the 'Big Freeze' or 'Big Chill'). It's an idea first proposed in 1851 by Lord Kelvin, a founding father of thermodynamics who was directly inspired by reading Sadi Carnot's book.

There is another possibility that would see a similar result in as little as 22 billion years. Astronomers call it the Big Rip. The rate at which the universe is expanding is currently increasing under the influence of a mysterious substance called dark energy, whose potency grows as the gap between galaxies increases. If this continues unchecked, it will become an unstoppable juggernaut. Stars, planets and even atoms will be torn apart and the universe will contain nothing more than shrapnel. It's a sobering thought.

Even if it's the Big Rip over the Big Chill, the universe will still have been around for tens of billions of years. The sheer longevity of space is down to the curious fact that the universe born in the Big Bang appears to have been in a highly ordered, low-entropy state. The entropy of the universe 13.8 billion years ago was a quadrillion times less than it is today. Had things been more chaotic, the cosmos would have started a lot further down the road towards its demise and there may not have been sufficient time to spawn planets and people at all. Exactly why the universe started out with very low entropy is one of the most perplexing mysteries in astronomy, standing unsolved for well over half a century.

If we figure it out one day, then time's arrow won't just tell us why we experience the past before the future; it could also tell us why we're here to experience time at all.

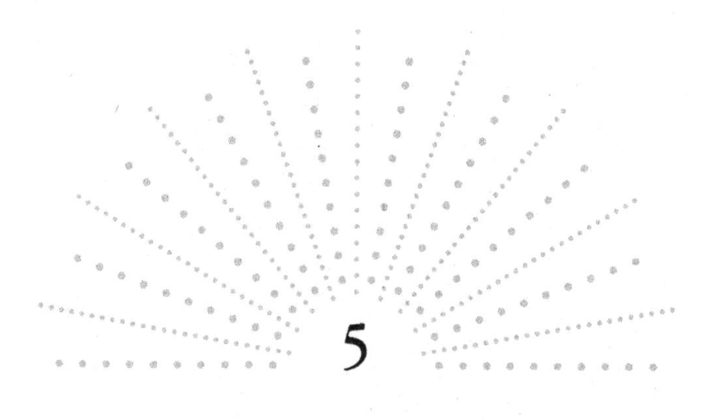

5

Space and Time Aren't as Different as They Seem

Arthur Eddington did more for our understanding of time than merely referring to its direction as an arrow. In 1919, the astronomer travelled to the tiny island of Príncipe off the west coast of Africa, thousands of kilometres from his birthplace in Cumbria, England. Eddington went all that way to observe a solar eclipse, a spectacularly beautiful and scientifically invaluable event that sees the Moon block out the Sun. With the First World War barely over, the Englishman was about to catapult a German into a level of celebrity scarcely matched before or since.

That man was Albert Einstein. Born in 1879, Einstein had a so-called miracle year in 1905 in which he published no fewer than three revolutionary papers, including one that would go on to win him the Nobel Prize in Physics in 1921. Another introduced the ideas that would later be encapsulated in his now world-famous equation, $E = mc^2$. This was all the more remarkable given that he was working in the Swiss patent office at the time, rather than in the hallowed halls of a university. Sometimes it takes an outsider's perspective to see what others cannot. In 1908, Einstein moved to the University of Bern and in that same year one of his former maths teachers made another important breakthrough (the very same teacher who once described Einstein as a 'lazy dog' who 'never bothered about mathematics at all'). His name was Hermann Minkowski and he used his former pupil's work to propose that, despite appearances, space and time are not separate entities.

Our experience of the world teaches us that they are. For starters, you have a fair amount of freedom in space. You can change direction, for example. Move one way, stop and then return the way you came. You can also change the speed at which you travel through space. You can sit still, walk, run, travel in a car, a plane or a rocket. Time, it seems, is not like that. We're stuck travelling in a single, unchanging direction from past to future – Eddington's arrow of time – and always at the same rate. It's no wonder we consider time and space as disconnected, yet Minkowski argued that they are intimately interwoven into a continuous fabric that pervades the universe. One that forms a stage on which all cosmic events play out. He named it simply by sticking the two words together: *spacetime*. Many of the remarkable and mind-expanding ideas you're going to encounter in the pages ahead hinge on this notion.

In 1915, Einstein published his general theory of relativity, which contained yet more controversial ideas, calling over two hundred years of scientific orthodoxy into question. It takes a bold person to claim that Isaac Newton – the godfather of physics – was wrong, but that's exactly what Einstein suggested. At the heart of the issue was the way we think about gravity. Newton famously thought of it as a force. Gravity pulls apples to the ground and it heaves heavy planets around the Sun. Yet Einstein thought differently, arguing that there was actually no pulling taking place at all. Instead gravity was simply the result of curved spacetime.

There's a frequently used analogy to help picture this. Imagine a bed sheet held tightly at the four corners to represent the fabric of spacetime, with a bowling ball placed in the centre to stand in for the Sun. The bowling ball creates a dip in the sheet and if you add in a tennis ball you can roll it around the rim of the dip, making it orbit the bowling ball just like the Earth circles the Sun. Crucially there is no Newtonian force of attraction between the two balls. The smaller one isn't being pulled by the larger one, it's just following the curved path it creates in the sheet. This was Einstein's hunch about how gravity really works. It was an extraordinary claim, one that required extraordinary evidence to back it up. That's exactly what Eddington was in Africa to find.

Both Newton and Einstein predicted that the Sun's gravity bends distant starlight around it. This has the weird consequence that a star sitting directly behind the Sun can instead be seen alongside it. The two men disagreed on how far the displaced star would appear from the Sun, which would provide an excellent way to see who was right. The catch? You can't normally see stars close to the Sun because it is so incredibly bright. During a solar eclipse, however, the Moon blocks out its glare and the stars close by become visible. Eddington travelled to Príncipe in 1919 to photograph the eclipse in the hope of measuring the positions of the stars close to the Sun. Despite heavy, intermittent cloud cover, he was able to take two revolutionary images. The stars were exactly where Einstein said they'd be. Newton was wrong. When the results were made public later that year, it was front-page news. 'New Theory of the Universe: Newtonian Ideas Overthrown', said *The Times*. The *New York Times* ran with 'Lights all Askew in the Heavens: Men of Science More or

Less Agog'. Such is the persistence of our everyday experience that many simply couldn't believe that space and time weren't separate. Instead, it turns out, they are two sides of the same coin.

In the century or so since Einstein first hit the headlines, physicists and astronomers have sought further evidence to back up his claims. One way to do this is to look for a larger-scale version of the Sun bending starlight in an effect called gravitational lensing. It relies on one galaxy sitting directly in front of another as seen from Earth. A reminder: a galaxy is a huge collection of hundreds of billions of stars. The foreground galaxy warps spacetime to such an extent that the light from the background galaxy is bent around it, much like a lens bends light to a focus. This forms a circle of light called an Einstein ring. You can see a similar effect if you look at an object through the curved base of a wine glass. The size of the ring depends on how much spacetime is curved, which in turn depends on the amount of stuff in the foreground galaxy. Astronomers consistently find that the rings match the size predicted by Einstein's theory. Another big green tick for the idea of space and time being intertwined.

Einstein also predicted that events taking place

within spacetime create waves that move outwards like ripples on a pond after a stone enters the water. Physicists call them gravitational waves and they were detected for the first time back in 2015, exactly 100 years after Einstein first published his landmark theory. The experiment used to make this historic discovery is called the Laser Interferometer Gravitational-Wave Observatory (LIGO) and it's made up of two machines working in tandem some 3,000 kilometres (1,864 miles) apart in the USA. Each detector consists of two 4-kilometre (2.49 -mile) -long arms at right angles to each other. A laser beam is fired down each arm towards a mirror at the end of the tunnel, which reflects it back to the start. As the arms are identical in length, the laser beams usually make it home together. However, that changes if a gravitational wave passes through during the lasers' passage. Gravitational waves wrinkle spacetime itself, temporarily changing the length of one of the arms. The lasers no longer arrive back at the start at the same time. It was such an important milestone that the scientists behind it won the Nobel Prize in Physics just two years later – one of the shortest gaps between discovery and accolade in the history of the Nobels.

Then, in April 2019, came yet another moment for the history books. Astronomers using the Event Horizon Telescope released the first ever photograph of a black hole. Black holes are gravitational giants: places that warp the fabric of spacetime to such a degree that all paths out are bent around and lead straight back in. It's why nothing can escape from a black hole. Missing light that has been gobbled up creates a shadow close to the black hole and Einstein's general theory of relativity makes very strong predictions about the size and shape it should take. In order to photograph the black hole and its shadow, astronomers used an array of telescopes spread across the planet. They generated so much data that it had to be loaded on to hard drives and flown by courier back to the project's headquarters. Remarkably, that was quicker than trying to send it all over the internet. Sure enough, the resulting photograph showed a perfect match with Einstein's predictions.

It's due to this avalanche of recent evidence that we're more confident than ever in the idea of spacetime and of its ability to be bent, shaped and distorted. Einstein's theory has passed every test thrown at it with flying colours. Space and time are so interconnected that what affects one, affects the other. If space can be bent, then so can time. That opens the door to the intriguing possibility of travelling not just through space, but through time as well . . .

6

Time Travellers Walk Among Us

Gennady Padalka was eleven years old when Neil Armstrong first set foot on the Moon. Like many children of his generation, he started dreaming of the stars. After an exemplary career in the Russian Air Force, he joined the Soviet cosmonaut programme five months before the fall of the Berlin Wall in 1989. Today Padalka is a space veteran, having flown on five different missions, conducted ten spacewalks and spent a record 879 days orbiting the Earth. He's also the greatest time traveller in human history.

The pages of science fiction are full of tales of time travellers, people who have fast-forwarded through time and skipped into the future. In the second instalment of the *Back to the Future* films, Doc Brown and Marty McFly hop from 1985 to 2015, encountering a futuristic world of hoverboards and flying cars. Sneaking a peek at tomorrow is a mesmerising prospect, but very few people realise that time travellers already walk among us. Gennady Padalka is far from a household name, despite being the closest thing we've ever had to Marty McFly.

Real-life time travel to the future hinges on an

effect known as time dilation. Imagine that you're running a 100-metre race with the eight-time Olympic gold medallist and world record holder Usain Bolt. You'd hardly be surprised if he reached the finishing line before you. After all, he has the ability to move through space faster than you (space as in physical space, not outer space). Except that the pair of you aren't really racing through space at all. You are racing through space*time* – the fabric of the universe we discussed in the last chapter. Space and time are so intimately linked that Bolt doesn't just travel through space faster than you, he also travels through time faster than you. By reaching the finishing line first, he also reaches the future first. The difference in speed between you and Bolt – although it may seem stark at the time – is so miniscule that the amount by which he beats you to tomorrow is tiny. You simply don't notice.

With Padalka it's a slightly different story. He spent time aboard both *Mir* and the International Space Station – orbital outposts that career around the Earth at 27,500 kilometres per hour (17,000 miles per hour). Padalka hurtled through spacetime much faster than those of us left on the ground for a total of 879 days. In doing so, he time travelled into the future by 0.025 seconds, more than anyone else in history. The cosmonaut is also a chrononaut.

Astronauts like Padalka are our most successful time travellers, but they aren't the only people doing it. *You* are also a time traveller. Time dilation means that every time you speed up you reach the future faster than if you were stationary. A return flight from London to New York, for example, will see you time travel into the future by 19 billionths of a second (19 nanoseconds). Even something as simple as walking does it. The steps you take over your lifetime will see you travel three nanoseconds into the future compared to just sitting still. You may be thinking that none of these amounts sound particularly impressive. Yet it shows that time travel to the future is more than possible. It's so simple that you've even done it yourself without Doc Brown or a DeLorean in sight. All we have to do to make it more interesting – more worthy of science fiction – is to travel through spacetime more quickly.

Often when people first hear about time dilation they are highly sceptical. It rails against our everyday experience of time to such a degree that they just don't buy it. However, you lean heavily on time dilation every time you use an indispensable tool of modern life: GPS. The Global Positioning System works thanks to a fleet of satellites orbiting the Earth. If you use the maps app on your phone, it will receive signals from these satellites. The quicker a signal arrives, the nearer you are to that satellite. By using several satellites at once, your phone can accurately pinpoint your location. For this to work, however, the satellites need atomic clocks on board to time how long the signals take to arrive. The trouble is that those machines are whizzing around the Earth at about half the speed of the astronauts aboard the International Space Station. Thanks to time dilation, the clocks on the satellites get out of sync with the clock on your phone. We have to adjust the GPS atomic clocks to bring them back in line. If we didn't correct for the fact that they have been time travelling, the whole system would quickly become useless. Left uncorrected for a single day, the little blue dot that tells you where you are would be out by some 10 kilometres (6.2 miles).

It's surprising that so few people know of this sure-fire route to time-travel success. After all, we've had evidence of its effects since the 1940s thanks to tiny sub-atomic particles called muons. Our planet is constantly bombarded by high-energy particles from space. Physicists call them cosmic rays. When a cosmic ray strikes an atom in our atmosphere it creates a shower of other particles, including muons. This is happening above your head right now to such an extent that 10,000 muons reach each square metre (11 square feet) of the Earth's surface every minute.

Scientists were initially puzzled by the sheer number of muons making it down here. Muons decay, just like the radioactive atoms we encountered in Chapter 2. Their half-life is incredibly short – only 1.56 millionths of a second. The muons seemed to be taking more than one hundred half-lives to reach the ground. With each half-life cutting the number of muons by 50 per cent, that's more than enough time for all but a few of them to decay and disappear. So how come so many muons arrive unchanged? The mystery was solved by taking time dilation into account and the fact that the muons travel at close to the speed of light. Considerably less time passes for them than it does for us, meaning that from their perspective only a few half-lives elapse en route to ground and a substantial fraction do not decay. Once you factor in time dilation, the number of muons reaching sea level tallies perfectly with predictions.

If humans could move even half as quickly as muons, our ability to travel in time would no longer be restricted to tiny fractions of a second, but measured in days, months and even years. Get a bit closer to their speed and we're talking decades. One way to illustrate the incredible potential of time dilation is to imagine a set of identical twins. Born just minutes apart in the same hospital, they grow up to choose different career paths. One becomes a doctor, the other an astronaut. When they are both forty the astronaut leaves Earth to command a daring mission to another star system. Rocket technology has improved so much that she can travel there and back at 90 per cent of the speed of light. Almost as fast as a muon. To her, the whole journey takes a little under twenty years and she returns just in time to celebrate her sixtieth birthday. Her twin remembers her own sixtieth birthday party well – it was twenty-five years ago. More time has passed for the twin left behind because she was travelling through spacetime more slowly. She's now eighty-five and has her great-grandchildren at her feet, telling them the story of her astronaut sister who has time travelled a quarter of a century into the future.

Why stop there? Let's say you travel through the

galaxy on a big loop at 99.9999 per cent of the speed of light – even faster than a muon – for ten years. You return a decade older, but seven *thousand* years will have flown by on Earth while you were away. You've swapped the twenty-first century for the ninety-first, experiencing life far beyond your lifetime. That's the kind of time travel people are usually interested in and there is nothing in the laws of physics to forbid it. We just have to learn how to travel faster and for longer than Padalka has so far.

Time travel this extreme is not without its problems, though. Imagine transporting someone living seven thousand years ago – an inhabitant of the Stone Age – to the middle of Times Square in modern-day Manhattan. That's one hell of a culture shock. Will humans even still exist that far into the future? If they do, will they remember who you are? If you put a pound in a bank account before you set off would you get seven thousand years of interest? All these issues and more will one day have to be considered and ironed out if we're to regularly travel through time by such large amounts.

There is an even bigger issue: time dilation is a one-way ticket. Unless you can come up with a completely separate way to travel backwards through time, returning to where – or when – you came from will be impossible. You're not overturning the arrow of time, merely hitting the fast-forward button and reaching the future faster. You can't use time dilation on its own to go the other way. Still, being marooned in the future is a price many would pay to see a world far beyond tomorrow. Would you?

7

Your Feet Are Younger Than Your Head

As the life of Peter Capaldi's version of the Doctor draws to a close, he finds himself aboard a most mysterious spaceship. The Time Lord and his companions encounter a solitary crew member who tells them that his 400-mile (640-kilometre) -long craft is stuck in the gravitational clutches of a black hole. Several of his colleagues went to reverse the engines in an attempt to power away from its grasp and haven't been seen since. Two days ago the vessel only had fifty inhabitants, but a quick life scan reveals that there are now thousands of souls on board. Surely invaders must have hijacked the already stricken ship? The ancient time-travelling Gallifreyan knows better. He realises that another form of time dilation is at play, one that doesn't involve travelling at high speeds. Gravity is affecting the rate at which time passes and the new people on board really are *new* people. They're the descendants of the original crew. A few days may have passed on the bridge, but a thousand years have elapsed at the other end of the ship.

To understand this gravitational time dilation, it pays to return to the analogy of the bed sheet and the bowling ball from Chapter 5. The bed sheet represents spacetime – the fabric that pervades the universe. Place the bowling ball in the centre to represent a massive object, like the Sun, and the sheet sags in the middle. Physicists call this depression a gravitational well. Spacetime is more curved and distorted inside a gravitational well than it is outside. Now, let's also imagine that we have a clock made from a beam of light bouncing between two mirrors. The clock 'ticks' every time the light hits a mirror. Outside a gravitational well, the light simply bounces up and down in a straight line between the mirrors. The deeper you go into a well, the more curved the spacetime between the mirrors becomes. The light follows this curved path, taking longer to travel between the mirrors and increasing the gap between ticks. Someone inside the well will experience less frequent ticks and so will say that less time has passed compared to somebody outside who sees more ticks. In the *Doctor Who* episode, one end of the spaceship was much further into the black hole's gravitational well than the other, with time passing at dramatically different rates as a result.

The same effect has some pretty bizarre consequences in the real world. Your feet are generally closer to the ground than your face, meaning they're deeper inside Earth's gravitational well. Less time passes for your ankles than your eyes. Your toes are time travellers, beating your tongue to the future by around half a millionth of a second over a typical eighty-year lifespan. Our heads are also travelling into the future at different rates depending on how tall we are. All else being equal, the brain of someone who is 1.5 metres (5 feet) tall reaches the future faster than the brain of someone who is 1.8 metres (6 feet) tall.

Where you live makes an even bigger difference. At an altitude of 5,000 metres (16,400 feet), the Peruvian town of La Rinconada in the Andes is the highest permanent human settlement in the world. Spend eighty years there – a little further outside the Earth's gravitational well – and you'll age 0.0025 seconds more than someone who whiled away their days at sea level. Physicists have even demonstrated that an atomic clock kept on a higher shelf in a laboratory will tick faster than another on a shelf just 30 centimetres (12 inches) lower.

This gravitational time dilation also affects our GPS satellites, which typically orbit 20,000 kilometres (12,400 miles) above the ground. That high up in Earth's gravitational well time passes more quickly by 45 millionths of a second each day. However, we need to take *both* forms of time dilation into account if we're to accurately correct the atomic clocks on board. Their speed sees time run slower by 7 millionths of a second daily. So the GPS clocks are altered by 38 millionths of a second per day (45 minus 7) to bring them back in line with ground-based clocks. At their altitude, gravitational time dilation is a more important effect than time dilation due to speed. For astronauts aboard the International Space Station – which only orbits 400 kilometres (249 miles) up and travels faster – it is the other way around.

Like the puny amounts of time travel achieved by Gennady Paldalka, these levels of gravitational time dilation are hardly going to set anyone's world on fire. Yet it is a further reminder that time is malleable – it can be bent and shaped depending on the circumstances. In the last chapter we had to travel much faster in order for more interesting time travel to kick in. This time, we need to find a much deeper gravitational well. Perhaps on Jupiter? It's the biggest planet in the solar system and has a gravitational field that's around 2.5 times stronger than Earth's. Unfortunately, even that doesn't do much to slow time. We're talking about a difference of less than minute over the average human lifetime. This is why writers – like those behind the *Doctor Who* episode – regularly turn to black holes. The biggest are a quadrillion times heavier than the Earth and their deep gravitational wells provide a meaningful amount of time dilation for science fiction.

Perhaps the most famous recent example of this is the 2014 Christopher Nolan blockbuster *Interstellar.* (If you haven't seen it yet and want to avoid spoilers, it's perhaps best to skip ahead to the next para-graph now.) It is set in a dystopian future in which humanity's existence on Earth is in jeopardy. A team of astronauts led by Cooper, played by Matthew McConaughey, go off in search of a new home. In order to get there, they use a shortcut through space called a wormhole (we'll come back to those in Chapter 9). They emerge in a distant galaxy close to a giant black hole called Gargantua, which is orbited by some potentially suitable planets. The astronauts are so deep inside Gargantua's gravitational well that a single hour spent on one of its planets is equivalent to seven years back on Earth. When Cooper does eventually return home, the ten-year-old daughter he left behind is now older than he is.

Could we actually do this? We know that at the heart of our Milky Way galaxy is a supermassive black hole called Sagittarius A* (pronounced 'A star'). It's about twenty-five times less massive than Gargantua in *Interstellar*, but still nearly three trillion times more massive than Earth. Suffice to say it has a very deep gravitational well. If you spent six and a half years hanging out 10 kilometres (6.2 miles) from the edge of the black hole, then 7,000 years will have passed on Earth. That's the same hop into the future achieved by whizzing around space at 99.9999 per cent of the speed of light for ten years in the last chapter. The catch is that in order to escape from the black hole's gravity and get home you'd still need to travel very close to the speed of light.

If your maximum speed was only half the speed of light, you could get within 50 million kilometres (31 million miles) of the black hole – a little closer than Mercury is to the Sun – and still be able to escape. A hundred days will pass on Earth for every 87 days that go by for you. You'd have to be gone for 67 years just to skip 10 years into the future (77 years will have passed on Earth). Significant time travel to the future is certainly possible, but it's definitely not easy. This is all without factoring in actually getting to the black hole and back – an unenviable 55,000-light-year round trip.

Time travelling using black holes is also fraught with danger. What if you misjudge the situation and venture too close like the crew in the *Doctor Who* episode? The edge of a black hole is called the event horizon. Cross it and the speed you need to travel to escape from the black hole exceeds the speed of light. Given that light is the fastest thing in the universe, that's impossible and there's no way out. No return home to see the far-flung future.

You wouldn't notice anything special happening as you crossed the event horizon, but someone watching your hapless mission from the outside would. They'd see time appearing to run ever more slowly for you as you approach the point of no return. At the event horizon itself they'll see your time grind to a halt. Your image will appear freeze-framed on the event horizon, a final portrait of your fate slowly fading from view.

Crossing an event horizon doesn't actually stop

time; it just appears that way to an observer far from the black hole. Once you're over the event horizon, no light can reach the onlooker and they can no longer see you. For you, however, time is proceeding normally. As we're about to see, whether or not it is possible to really stop time stone dead depends on what's at the centre of a black hole . . .

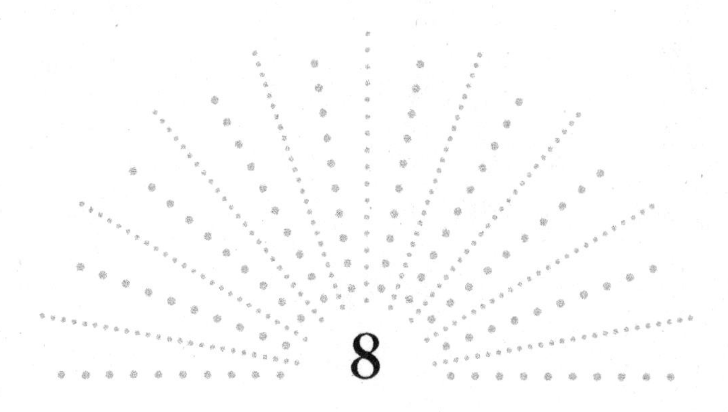

8

Time Can Be Stopped (Maybe)

John Wheeler received a postcard from his brother Joe in the summer of 1944. Just two words were scrawled on the back: 'Hurry up'. Joe was fighting in Italy during the Second World War, while John was working on the development of the atomic bomb at Los Alamos as part of the famed Manhattan Project. Although the bomb was a secret, Joe guessed what John was up to, given the nature of his work on nuclear physics before the war. He could see the war-ending potential of such a destructive weapon and was urging his brother to get a move on. The bomb wasn't finished for another year and Joe died in battle in October 1944, something John never entirely came to terms with.

After the war, John, known by everyone as Johnny, returned to Princeton University where he became Einstein's colleague and collaborator. He picked up the great man's work on general relativity, teaching the first ever graduate course on what was still a largely overlooked subject. Wheeler and his students returned to a scientific paper that had been published on 1 September 1939, the very same day Germany invaded Poland. The ensuing conflict meant that eyes had largely been focused elsewhere. The paper was called 'On Continued Gravitational Contraction' and it was co-authored by Robert Oppenheimer, the father of the atomic bomb and Wheeler's colleague at Los Alamos. It was the first scientific paper to describe the formation of an object Wheeler would later call a black hole[4].

One of the puzzles Wheeler worked on with his students was what happens to you if you're unfortunate enough to plunge over the event horizon of a black hole. Let's say you go in feet first. The gravity of the black hole will pull harder on your toes than on your head. This is true of Earth's gravity too, but the difference is beyond tiny. Inside a black hole it is significant, enough to stretch you out into long, thin strips of human spaghetti. You get torn apart lengthways in a process genuinely called spaghettification. The pasta of your constituent parts will arrive at the centre of the black hole anywhere

---

4 In a weird twist of history, the same journal issue carried a paper co-authored by Wheeler entitled 'The Mechanism of Nuclear Fission', describing the physics that Oppenheimer would later use to build the atomic bomb. Therefore details of the process that would help end the war were published on the very day it started.

between a fraction of a microsecond and almost a day later depending on the size of the black hole. But arrive where exactly? What is at the centre of a black hole and what happens to time there?

To answer that, we need to look at how a black hole forms in the first place. As Oppenheimer stated in 1939, it's from continued gravitational contraction. A star in the prime of its life exists in a delicate balance between the relentless heave of gravity trying to collapse the star and the starlight being produced in its core fighting in the other direction. Eventually the star exhausts its fuel reserves and starlight production stops. At that point gravity wins and the star collapses. This usually results in stellar material contracting into a really compact object. A neutron star, for example – the result of a mid-size star collapsing – is the size of a city but contains half a star's worth of stuff. A single spoonful of neutron star material weighs more than Mount Everest, or one hundred times the total mass of every human who has ever lived. You'd have to travel at nearly half the speed of light in order to successfully climb out of a neutron star's deep gravitational well.

For the most massive stars – those over thirty

times the mass of the Sun – they keep on collapsing instead of bottoming out as a neutron star. The gravitational well becomes so deep that the speed you'd need to travel to escape exceeds the speed of light and, voila: we have the cosmic trapdoor that is a black hole. The star keeps on contracting into an infinitely small, infinitely dense point called a singularity. According to Einstein's general theory of relativity, the more mass you have in a small area, the more spacetime curves in on itself. At a singularity, spacetime is wrapped up into an infinitely small speck and the usual concepts of space and time curl out of existence. 'Every black hole brings an end to time,' Wheeler noted in his autobiography, *Geons, Black Holes and Quantum Foam*. Hit a singularity and you are summarily erased from the universe (or at least your spaghettified bits are).

Many physicists don't buy the idea of a singularity. They see it as a placeholder for our ignorance, arguing that at such tiny scales it's the general theory of relativity that breaks down instead of time and space. Part of the issue is that in the picture described above there is no mention of quantum physics – the laws that govern atoms and sub-atomic particles. Normally the two theories are easily separated. You don't need to worry about quantum effects on the scale of planets and stars; you shouldn't concern yourself with gravity and spacetime when you're dealing with tiny atoms. Except a black hole muddies the waters. Matter starts off the size of a star and ends up being crammed into a space smaller than an atom. To really understand what happens to time at the centre of a black hole you probably need to combine the two into a single theory that is widely known as quantum gravity.

If only it was that easy. Quantum physics and the general theory of relativity go together like pineapple and pizza. They don't. Legions of physicists have tried for decades to get the pesky equations to blend nicely together. A theory of quantum gravity is perhaps the most sought-after prize in all of physics. It's also the most elusive. At the heart of the discord is a fundamental disagreement about smoothness. General relativity invokes the smooth, continuous fabric of spacetime. Quantum physics says everything is chunky. Light, for example, comes in chunks of energy called photons (another discovery of Einstein's and the one he won the Nobel Prize for). There's every reason to think that combining the two should be possible. Take this book. It is made of atoms, which are governed by the rules of quantum physics. Yet if you drop it, it will fall to the floor as described by the general theory of relativity. It would be bonkers to always need two theories to explain one event.

There *are* ways to combine the theories, but to do so physicists always have to add something extra. String theory, for example – the same field Sheldon Cooper studies in TV's *The Big Bang Theory* – invokes up to seven extra dimensions in addition to the three spatial dimensions and one time dimension we experience. For us not to notice them, they are said – rather conveniently – to be too small to see. There are also a huge number of possible versions of string theory – '1' followed by a whopping 500 zeroes. For comparison, the number of atoms in the entire observable universe is 'only' '1' followed by 80 zeroes. We don't know which one, if any, of the countless versions of string theory applies to our universe.

A rival theory – loop quantum gravity – says that spacetime isn't a smooth fabric at all, but chunky like everything in the quantum world. If we could zoom in and see spacetime on the smallest scales, we'd see it was made of a series of 'stitches', just like the bed sheet we've used as an analogy for spacetime in previous chapters. Or think of spacetime as a seemingly smooth image on a computer screen. Zoom in and you'll see that it's actually made up of tiny squares called pixels. Atom smashers like the Large Hadron Collider (LHC) at CERN in Switzerland are probing smaller and smaller scales, but to test loop quantum gravity on Earth you'd need a machine 1,000 trillion times more powerful than the LHC. The universe could help instead. Starlight travelling to us through spacetime could be affected by its underlying structure. The effect of each stitch would be tiny, but the sheer number of stitches that light encounters on the way could show up as something measurable.

The trouble is that there's no concrete evidence yet that any of these additions are actually present in the real universe. Physicists can write down equations, just like comic book writers can pen stories about superheroes. But that doesn't make string theory any more real than Spider-Man. Unless we can successfully identify a theory of quantum gravity, it may not be possible to say with any certainty whether time really does stop in the belly of a black hole. We can hardly send anyone in for a closer look.

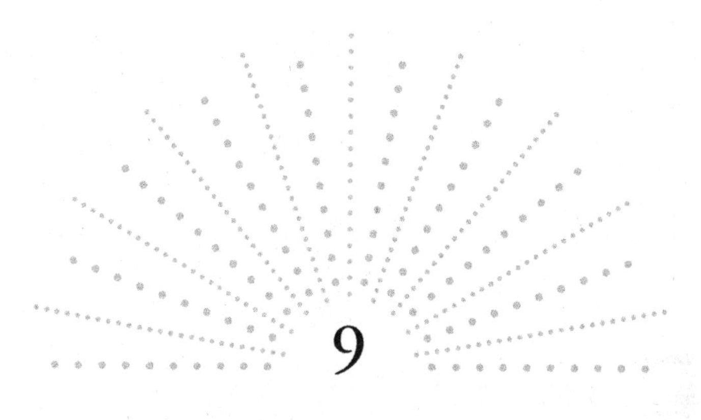

9

Can You Kill Hitler?

When it comes to time travel, there is something people seem to want more than anything else: to travel backwards in time. Perversely the past is more alluring because we already know what's happened. We can skip over the boring bits and pick and choose our moments. Maybe you want to dine with Henry VIII or become Boudicca's drinking buddy. It could be that you'd like to try to undo past mistakes or see loved ones no longer with us. The Black Death and the COVID-19 pandemic are perhaps best avoided. Travelling into the future is more of a shot in the dark – who knows where we'll end up or what sights await us?

You don't have to spend long digging into the possibility of backwards time travel before the name of one man inevitably crops up: Adolf Hitler. In 2015, a poll by the *New York Times* magazine asked its readers whether they would go back in time and kill a baby Hitler if they had the opportunity. The result: 42 per cent of people said they would, with 30 per cent saying no and the rest undecided. But could you actually do it even if you wanted to?

The good news is that physicists *have* come up

with a way to travel backwards in time. A blueprint for a time machine that breaks no known laws of physics. All you need is a wormhole – the same shortcut through space used by the astronauts in *Interstellar* and named as such by John Wheeler. These enigmatic objects remain theoretical, but they are consistent with Einstein's general theory of relativity and, so far, that theory has passed every test thrown at it. As we saw in Chapter 5, black holes were once equally far-fetched objects, but now they are a mainstay of mainstream science.

The easiest way to picture a wormhole is to start with an ordinary sheet of paper that represents spacetime. Earth is at one end of the paper and the nearest star to us after the Sun – Proxima Centauri – is at the other end. To travel between these two locations you would usually have to traverse the length of the paper, which is an unenviable 40-trillion-kilometre (25 trillion mile) or 4.2-light-year trek. However, we've already seen in earlier chapters that spacetime can be manipulated and curved. If you fold the paper in half, suddenly your destination is considerably closer. A tunnel between the paper's edges would allow you to get there in no time at all. That tunnel is a wormhole.

Now for the time-travel bit. First, attach the Earth end of the wormhole to a rocket and whizz it around space at close to the speed of light, returning to the Earth just over five years later. Five years may have passed on Earth, but the rocket's speed means that time has been running at a slower rate inside the wormhole thanks to time dilation (see Chapter 6). Only six months will have gone by at the other end. So if you now jump in the Earth end of the wormhole you will emerge at Proxima Centauri 4.5 years ago. You have successfully time travelled into the past and your place in the history books is assured. It may be a revolutionary feat, but it's not a particularly exciting one. Most people who want to time travel would rather return to a point in *Earth's* history. Fear not. All you have to do is get on another rocket and travel back home the long way round (not using the wormhole) at close to the speed of light. A little over 4.2 years will elapse on Earth during that 4.2 light year journey, meaning you arrive home months before you left Earth in the first place. Wait for those months to pass and you could watch the past version of yourself make history by jumping into the wormhole.

Here's an example:

Action	Date on Earth	Date at Proxima Centauri
Attach Earth end of wormhole to rocket	1 Jan 3000	1 Jan 3000
Return wormhole to Earth after loop around space	10 Jan 3005	2 July 3000 (time dilation)
Jump in Earth end of the wormhole	**10 Jan 3005**	2 July 3000
Leave Proxima Centauri to return to Earth	-	2 July 3000
Arrive back on Earth	**5 Oct 3004**	-

The big caveat with this kind of time machine is that you cannot use it to travel back to a point before you started building your time machine in the first place. The time at the other end of the wormhole will always be *after* you attach the wormhole to your rocket (but before you jump in). As we haven't done this yet, there's no killing Hitler, banqueting with Henry VIII or partying with Boudicca. That is unless an alien civilisation elsewhere in the universe already started setting up a wormhole time machine before these events happened. If they did, then you would be free to use it to attempt these feats.

Although theoretically achievable, time travel to the past makes a lot of physicists deeply uncomfortable because it throws up seemingly contradictory situations called paradoxes. What if, upon returning back to the Earth months before you left, you buy a gun and shoot your past self before they could jump into the wormhole? If your past self is dead, they couldn't have used the time machine. So, who travelled back in time to buy the gun and shoot them? This strange state of affairs often goes by the name of the grandfather paradox. If you go back in time and kill your grandfather when he was a child, then you can never have been born, let alone travel back in time to murder him. This is beautifully illustrated in *Back to the Future*. Marty McFly's mother starts falling for him at the high-school prom, meaning she never gets together with Marty's father. As this means Marty could never have been born, he begins to fade away. Luckily he rectifies the situation before he disappears in a puff of time-travel logic.

What about that original, seemingly noble idea of

killing baby Hitler? If you execute him as a child, then he never grows up to become the leader of the Nazis. Which I'm guessing was your plan all along. But then he never does those terrible things, you never learn about him at school and therefore you can never decide it's a good idea to go back in time and kill him. What reason have you got for wanting to shoot an otherwise ordinary little Austrian boy?

In both the case of your grandfather and Hitler you would *have* to fail in the assassination attempt you're about to make for you to be there at all. Maybe someone sneezes nearby, forcing you to miss. Perhaps the gun jams. This is known as the Novikov self-consistency principle after Russian physicist Igor Novikov. To get around the paradox, the action that would cause it to happen cannot occur. The trouble is that this makes your future predetermined and erodes any notion of free will. No matter how hard you try to kill Hitler, you can't.

Your free will is also called into question by the so-called predestination paradox. Let's say you go back in time to investigate the cause of a house fire, but while you're there you accidentally knock over a candlestick setting the curtains ablaze. As you flee to safety, the penny drops: *you* started the fire. If you don't knock over the candlestick, then the fire never happens and you can't travel back in time to investigate it. So, knock over the candlestick you must. Your future is fixed and there is nothing you can do to stop it. It's either that or time travel to the past is impossible. You can't have it both ways.

As if that wasn't enough to contend with, the bootstrap paradox adds yet another layer of trouble. Let's imagine that you go back in time to 1900 with a copy of Einstein's general theory of relativity because you want to get it autographed by the man himself. Crucially the theory wasn't published until 1915. Einstein is so impressed by what he sees that he steals it from you and publishes it fifteen years later as his own work. Who actually came up with the theory? It can't be Einstein – he pinched it from you. And it can't be you, because you got it from Einstein. It's no wonder David Tennant's Doctor referred to time as a 'a big ball of wibbly-wobbly, timey-wimey . . . stuff' in Doctor Who.[5]

You also ought to think very carefully about

---

5 If you want to see time-travel paradoxes taken to a mind-bending extreme, look up the short story '"– All You Zombies –"' by Robert A. Heinlein.

whether you really want to be the inventor of the first time machine at all. It will exist for all future time, unless it is destroyed or breaks at some stage. In all likelihood, a large number of people will want to use the time machine to travel back in time and meet its inventor. The moment you turn it on you'll be inundated with time-travel tourists from the future. You'd be so overwhelmed with requests for autographs and photographs that you would never get to use the time machine. It might even drive you to destroy the damn thing yourself. My advice? Be the second person to invent the time machine.

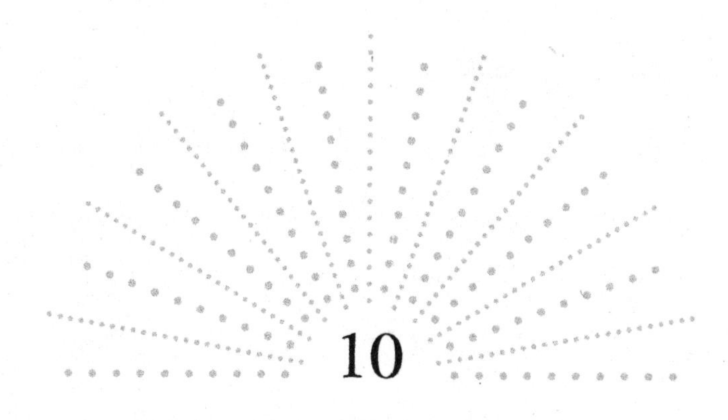

10

Time May Not Even Exist

Michele Besso was an Italian engineer. He was also Albert Einstein's best friend. They were students together in Zurich and it was Besso who helped Einstein get his famous job at the patent office in Bern. When Einstein was putting together his revolutionary theories of relativity, it was Besso who he turned to for a kindly ear. The two men died just a month apart in 1955; Besso first. One of the last letters Einstein ever wrote was to Besso's grieving family. Trying to offer a modicum of comfort, he said: 'He has departed from this strange world a little ahead of me. That means nothing. People like us, who believe in physics, know that the distinction between past, present, and future is only a stubbornly persistent illusion.'[6]

To see what Einstein was getting at, ask yourself a seemingly simple question: where does tomorrow *come from?* As we've seen on numerous occasions, Einstein's theories tell us that the universe is one enormous heap of spacetime. According to our best

---

6 William J. Hoye, *The Emergence of Eternal Life*, Cambridge: Cambridge University Press, 2013, p. 187.

current theories, the Big Bang created all of space and time (see Chapter 2). There's no process that can create more spacetime. So it cannot be that tomorrow doesn't exist yet because we've just said that you cannot create more time. It must exist already. Equally, the past can't magically disappear. If it did, spacetime would need to be constantly removed from the universe. Drive down a long road and the next bit of tarmac doesn't suddenly appear the moment you approach. Nor does the road vanish as soon as you've driven over it. Why should time be any different to space? After all, Einstein's already told us that they are two sides of the same coin.

This line of reasoning results in a staggering conclusion: the entirety of the past, present and future must be out there. Our entrenched sense that only 'now' is real is wrong. The universe already contains everything that has ever happened and everything that ever will. Somewhere out there you are being born and somewhere else you are dying. In yet other places, JFK is being shot, the World Trade Center towers are collapsing and your great-great great-grandchildren are laughing at the fact that you drove your own car. Everyone who has ever lived and everyone who will ever live co-exist, just in different parts of spacetime. Einstein was trying to comfort Besso's family by telling them that, in a very real way, Michele is still out there.

This idea is known as the block universe and I'll lay a bet that it makes you deeply uncomfortable. You're not alone. In some ways it should rankle because it jars so resolutely with the way we experience time. It *feels* wrong even if we can't quite put our finger on why. Yet in another way it shouldn't be so troubling. We're more than comfortable with the idea that other locations in space simultaneously exist alongside our current location. We don't say Mars doesn't exist just because we're currently on Earth. Equally, the region of spacetime known as tomorrow already exists even though we are currently located in another called today. It's perfectly possible for two people to be in different locations in space and both correctly identify as being 'here'. If space and time are intertwined then why should that suddenly change if you swap 'here' for 'now'?

Another way to look at it is by imagining the block universe as a book. If you could somehow look at the book from the outside, you would see that both the past and the future have already happened. The ending is already written. The book itself is static and unchanging – it just sits there like a block of concrete (which is where the name comes from). We, however, are stuck *inside* the book. As we move through the static pages one by one, we compare each new page – each new moment – to the last. The change that we inevitably experience gives us the *impression* that we're being carried along by a river of time that flows relentlessly from past to present to future. Yet we don't say space passes or flows. Time doesn't either. The flow of time is a figment of our imaginations.

If you can accept that time isn't really flowing and that yesterday, today and tomorrow are co-existing locations in spacetime, then some of the strange notions we encountered in earlier chapters actually make *more* sense. Each of us is taking a different path through the book of spacetime depending on how fast we're travelling. If I travel through spacetime faster than you, then I'll race through the pages at a greater pace and get to the end – the future – more quickly (time dilation). The faster I travel compared to you, the bigger the difference between us. You can also bend a page around in a loop so that it intersects with a previous page (time travel to the past). Yet you can't change what's already written there, which is why you can't kill either Hitler or your grandfather when they were children.

On a more philosophical level, the block universe theory says that the future is already out there waiting for you. You can't change your story any more than you can alter the end of this book by reading this page. If tomorrow isn't created by what happens today, then that notion you have of free will is as much of a mirage as the flow of time. The future already exists and is unchangeable. On the plus side, you needn't berate yourself for making bad decisions. You were always going to make them and the unwelcome resulting outcome was always inevitable. Some might call it fate. On the downside, why strive hard to achieve anything when there's nothing you can do to improve your future lot? Personally, I wouldn't sweat it. It's not like you can suddenly change tack after reading this book. If you do, then you were always going to make that decision anyway. Welcome to actualism – the idea that only what actually happens ever could have happened. There is only one possible, predetermined future.

If that has sent you into an existential tailspin, know that not every physicist and philosopher agrees with the block universe interpretation of time. It is currently, however, the mostly widely held view. At least time is a real entity in the block universe, even if the flow of time isn't. Some physicists want to go even further and dispense with time completely.

We saw in Chapter 8 that physicists currently have two different theories for explaining the way the universe operates. Einstein's general theory of relativity accounts for gravity and the very big, while quantum physics explains atoms and the very small. At what point do you need to switch from using one to the other? John Wheeler found himself pondering this very question in 1965 while at Raleigh-Durham International Airport in North Carolina. He had a two-hour stopover between flights and invited fellow physicist Bryce DeWitt to join him to spitball ideas. Between them they wrote down a quantum gravity equation that has gone down in physics folklore. Unsurprisingly it is called the Wheeler-DeWitt equation and there is no mention of time anywhere in sight.

Any successful theory of quantum gravity will be a more fundamental theory than either quantum physics or general relativity in the same way that general relativity is a more fundamental theory of gravity than Newtonian physics. No one uses relativity to send probes to Mars as Newtonian physics is a good enough approximation. Even though gravity isn't really a pull – it's the result of curved spacetime – most of the time it makes sense to imagine it that way. Equally, quantum physics and general relativity would be good approximations to a deeper theory of quantum gravity. The fact that the Wheeler-DeWitt equation – a potential equation for quantum gravity – can do without time entirely has led some physicists to question whether time really exists at all. Time could be a useful illusion in the same way that a Newtonian gravitational 'pull' is.

To see why, put two fingers on your wrist and take your pulse. You can feel the effects of your heart pumping, but it only really pumps on a large scale. Zoom down to individual heart cells and there's no pumping going on. That thing we call pumping only emerges when lots of cells operate in combination. Physicists would say that pumping is an *emergent* property of the system rather than a fundamental one. If the Wheeler-DeWitt equation is correct – and that's a big if – space and time could be emergent effects that we experience only at a large scale but that don't actually exist at all at a deeper level. A god-like observer, who could somehow see the cosmos from the outside, would see a static, unchanging and timeless universe. It would render time a mere illusion for those of us stuck inside the universe. It might turn out to be all in our heads. Ironically, only time will tell.

Epilogue:
A Time-travel Experiment

Dearest Time Travellers,

On February 26 2020 I gave a talk on time travel at the Ritz Cinema in Brixton, London, England. We were in Screen 3 and it took place between the hours of 18.30 and 20.00. It would be great if you could join us to tell us what time travelling is really like. So I know that it is you, bring a red umbrella, a ham sandwich and say the word 'Rumpelstiltskin'. Oh, and you'd better bring a copy of this book. I wasn't approached to write it until May 2020 and so I'll have no idea what you're banging on about. I'll try not to steal it from you and publish it as my own work . . .

Of course, if the block universe theory is correct, then I'll have already remembered it happening before I started writing. I was always going to write this book and you were always going to travel back in time to meet me. Then again, maybe what I do today does change the future. If someone ever uses this invitation, the moment I type these words my memory will instantly change from one with no time travellers present to one who remembers someone disconcertingly whispering 'Rumpelstiltskin' in my ear while brandishing a ham sandwich and a red umbrella.

I'm sad to report that I have no such memory. So either we never invent a way to time travel to the past or you've deemed me too unimportant to waste your machine on. How rude.

Safe travels,
Colin Stuart
October 2020

Acknowledgements

Any book is a collaboration between multiple people spread out across both space and time. So firstly I'd like to thank my teenage self for sticking with physics even though it wasn't always easy. I think you'd be pretty happy with the way things have turned out.

Ruth, my wife, deserves a huge amount of the credit for being both my anchor in tough times and my rudder in better times. I love sharing spacetime with you.

Thanks also to my nephew, Max, for his fearsome ten-year-old intellect. He's always a great person to bounce ideas off and grasped the implications of the bootstrap paradox faster than any adult I know. It wouldn't even surprise me if he built a time machine one day.

The people who work behind the scenes often don't get the credit they deserve, so thanks to my editor, Ru, and agent, James, for helping get this book to you.

And it's you I'd like to thank last. Thank you for taking the time to read about time – I hope you'll see your place in this crazy universe a little differently from here on in.